K-Math Workbook Grade 6

A Smart Way of Learning Math

Jaehwa Choi, Ph.D.
The George Washington University

Sunhee Kim, Ph.D.
Howard Community College

Kyongil Yoon, Ph.D.
Notre Dame of Maryland University

CAFA Lab

How to Use K-Math Workbook

All Practice Pages are fully organized by the Common Core State Standards (CCSS). By scanning each **Item's QR code** with any smart device, one can instantly access these additional features:

Detailed Solutions
Get the answer with detailed solution steps

Item Drills
Drill the item variants until fully mastering the item

More Practice
Practice other items until fully mastering each standard

By scanning the **Exam QR code** in the Exam Page, one can submit answers and get immediate feedback such as: grades, correct answers, solution steps, additional practices, and diagnosis reports.

K-Math Workbook, integrating Information and Communication Technologies (ICT) into a classical paper workbook, is designed to cope with these Common Core State Standards challenges:

- Help to articulate to parents, teachers, and the general public expectations using a simple but Concrete Application
- Provide aligned textbooks, digital media and curricula to international standards with a Workbook Solutions
- Help to implement an assessment system to measure student performance on standards via a Formative Assessment System
- Provide professional development opportunities for educators on identified needs and more efficient practices with a Teacher Friendly Application
- Help evaluate policy changes needed to help students and educators, so they can meet the standards through a Research Oriented Approach

For more information about K-Math Workbook visit http://CAFALab.com.

Table of Contents

Ratios and Proportional Relationships

6.RP

1. For every 56 minutes Tom studied, he took a rest for 13 minutes. What is the ratio of minutes spent studying to minutes spent resting?

2. Kristy counted the numbers of candies and chocolates after trick or treating. There were 320 candies and 200 chocolates. What is the ratio of the number of candies to the number of chocolates?

3. On a rectangle, the ratio of the width and the length is 12 : 5. If the length is 10 inches, what is the width in inches?

4. In 6th grade, there are 72 boys and 64 girls. What is the ratio of the number of boys to the number of students?

1. Megan counted how many shirts and pants she had. The ratio of the number of shirts to the number of pants is $5:2$. What unit rate is equivalent to this ratio?

2. The competition ratio for an exam is $3:1$. If 450 people took the exam, how many people passed the exam?

3. The unit rate of a cup of sugar for a cup of flour is $\frac{7}{8}$. If 24 cups of flour are needed to bake cookies, how many cups of sugar are needed?

4. The unit price for a pizza is $13. If John paid $91, how many pizzas did he buy?

1. Kristy ate cookies for snack, and found that there were 4 grams of sugar for every 80 grams. If she ate the whole bag which weights 240 grams, how man grams of sugar did she eat?

2. Ethan drove 60 mph to go home from his college. The distance between his college and his house is 300 miles. After he had been driving for 3.5 hours, how far is he to his destination? Answers in miles.

3. At Jenny's school, there were 500 students last year. If there are 20% fewer students this year, how many students are at her school this year?

4. What is the percentage of the shaded region?

1. The ratio of two values in each column in the table below is equivalent. What is the value of x?

5	7	9	10
25	x	45	50

2. What is the ratio of the data in the table?

5	7	9	11
40	56	72	88

3. What is the unit rate of values of x to values of y for the following graph?

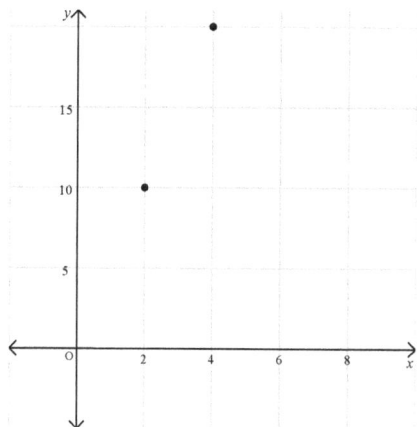

4. What is the unit rate of t to N?

t	N
3	12
10	40

1. The car runs 260 miles in 5 hours. If the driver keeps the same speed, how many miles can he drive in 7 hours?

2. Mr. Simon's art class has female students to male students ratio of 4 : 5. If Mr. Simon's art class has 16 girls, how many students are in his class?

3. If the price of 5 pounds of apples is $25, what is the price of 6 pounds of apples?

4. If it took 9 hours to mow 8 lawns, then at that rate, how many lawns could be mowed in 45 hours?

1. What is 25% of 26?

2. Kristy was organizing her books. There were 120 nonfiction books out of 240 books. What percent of nonfiction books does she have?

3. At a town meeting, 45% of attendees voted in favor of a new law. If the number of attendees is 300, how many people voted against new law?

4. Kristy went to a restaurant for lunch. The bill for her lunch was $14. If she left 19% tip, what was the total cost for lunch?

1. One gallon of orange juice costs $16. What is the price of one cup of orange juice?

2. At a county fair, 1 ounce bag of candies is $0.84. What is the cost for 3 lb bag of candies? Round to the nearest cent if necessary.

3. James was walking to a park that is 2 miles away. If he walked 110 yards per minutes, how many minutes would it take him to walk to the park?

4. A family of 5 went on a trip. They calculated the number of gallon size water bottle to purchase. They estimated 7 cups a day per person and their trip was 4 days long. How many gallon size water bottles did they have to purchase?

1. There are 1216 balls in a large bag. Out of these, 304 are red balls, 532 are blue balls, and the rest are yellow balls. What is the ratio of the number of blue balls to the number of yellow balls?

Ⓐ 1:4   Ⓑ 4:5

Ⓒ 7:16   Ⓓ 4:7

Ⓔ 7:5

2. Bob's family went to restaurant. They paid $66 for 11 burgers. What is the unit price of a burger?

Ⓐ $4   Ⓑ $55   Ⓒ $12

Ⓓ $6   Ⓔ $66

3. At one town meeting, 172 people attended and 68 people voted in favor of new law. What is the ratio of the number of votes against the law to the number of votes in favor of the law?

Ⓐ 17:26   Ⓑ 26:17

Ⓒ 17:43   Ⓓ 43:17

Ⓔ 26:43

4. Ian found that he grew 3% of his height for one year. If his height was 58 inches last year, what is his height in inches now?

Ⓐ 63.24   Ⓑ 59.74   Ⓒ 60.24

Ⓓ 61.24   Ⓔ 59.24

5. Ms. Emily's class has a female student to male student ratio of $3:4$. If the total number of students in Ms. Emily's class is 28, how many boys does she have?

Ⓐ 12   Ⓑ 14   Ⓒ 16

Ⓓ 18   Ⓔ 10

6. When each column in the table has the same ratio, what is the missing value in the table?

5	15
	27
20	60

Ⓐ 12   Ⓑ 10   Ⓒ 9

Ⓓ 8   Ⓔ 11

7. The ratio of marbles that Rachel and Megan have is $8:5$. If Rachel has 24 marbles, how many marbles does Megan have?

Ⓐ 39   Ⓑ 25   Ⓒ 6

Ⓓ 120   Ⓔ 15

8. In one elementary school, 35% of students are boys. If there are 154 boys, how many students are in the school?

Ⓐ 475   Ⓑ 405   Ⓒ 594

Ⓓ 286   Ⓔ 440

9. Rachel has 56 beads and Christine has 8 more beads than Rachel. What is the ratio of the number of beads that Rachel has to the number of beads that Christine has? Use the simplest form.

Ⓐ 7 : 8   Ⓑ 1 : 1   Ⓒ 7 : 1

Ⓓ 56 : 8   Ⓔ 56 : 64

10. If a car runs 13 miles in 10 minutes and keeps the same speed, how many hours would it take to run 312 miles?

Ⓐ 4   Ⓑ 1   Ⓒ 240

Ⓓ 10   Ⓔ 52

11. The ratio of the number of marbles that Jane has to the number of marbles that Kristy has is 6 : 5. If there are total 55 marbles, how many marbles does Kristy have?

Ⓐ 5   Ⓑ 25   Ⓒ 30

Ⓓ 55   Ⓔ 7

12. Joanne and Christine divided 70 clips into the ratio of 7 : 3, and Christine gave Joanne 7 clips. What is the simplest ratio of the number of clips that Joanne has to the number of clips that Christine has?

Ⓐ 7 : 3   Ⓑ 4 : 1   Ⓒ 1 : 1

Ⓓ 49 : 21   Ⓔ 56 : 14

13. At a zoo, dolphins and seals combined eat 70 pounds of food. If dolphins eat 4 times more than seals, how many pounds of food do seals eat?

Ⓐ 14   Ⓑ 55   Ⓒ 16

Ⓓ 15   Ⓔ 28

14. The perimeter of a rectangle is 130 inches and the ratio of the width to the length is $12.8 : 8$. What is the area of this rectangle in square inches?

Ⓐ 25   Ⓑ 40   Ⓒ 1040

Ⓓ 1000   Ⓔ 1025

15. At one community, 90% of students walk to school. If the number of students that walk to school is 90, what is the total number of students in that community?

Ⓐ 105   Ⓑ 110   Ⓒ 100

Ⓓ 115   Ⓔ 95

16. On a map, 1 inch represents 1750 feet. What is the scale on this map?

Ⓐ $21000 : 1$ Ⓑ $1750 : 1$ Ⓒ $1 : 1750$

Ⓓ $1 : 50$ Ⓔ $1 : 21000$

The Number System

6.NS

1. Christine's bakery uses $\frac{1}{7}$ lbs of chocolate each day. If they have $\frac{4}{7}$ lbs of chocolate, how many days will it last?

2. Kristy has $3\frac{5}{6}$ yards of string. For her project, she needed to cut string into $\frac{1}{18}$ yard lengths. How many pieces of string will she have?

3. After trick or treating, 5 kids measured the weights of their candies and that was $2\frac{1}{7}$. If they decided to share exactly same amount, how many pounds will each kid have?

4. What is the length in feet of a rectangle with width $\frac{2}{3}$ ft and area $\frac{8}{11}$ square ft?

1. Given a sequence: 8, 16, 24, 32 ... , with an increment of 8. Which number is _not_ in the sequence?

Ⓐ 297   Ⓑ 48   Ⓒ 136

Ⓓ 288   Ⓔ 144

2. What is the remainder when 173 is divided by 22?

3. Megan has 3.64 lbs of food for her cat. If her cat eats 0.91 lbs each day, how many days will the food last?

4. Emily borrowed a book from the library. The book has 259 pages and she wanted to finish the book in 7 days. If she planned to read the same number of pages each day, how many pages does she have to read each day?

1. Divide 2.1 by 7.

2. Emily gets $8 every week for her allowance. She spends $5.8 for snack each week and saves rest of her allowance. How much will she have at the end of 5 weeks?

3. At a party, Kristy ate 3 cookies and 6 cups of juice. When she checked the nutrition facts, she found that each cookie contains 3.5 grams of sugar and 2-cups of juice contains 3.6 grams of sugar. How many grams of sugar did she eat?

4. Kristy poured 2 lbs of flour in the container to bake cakes. Then she added 1.4 lbs of flour into the same container. How many pounds of flour is in the container?

Skill Practice: Find GCF and LCM of two whole numbers
(6.NS.B.4)

Name: _____

Date: _____

---

1. What is the greatest common factor of 48 and 64?

2. What is the least common multiple of 105 and 45?

---

3. Which of the following is equivalent to $30 + 45$?

Ⓐ $15(2 + 15)$ Ⓑ $15(2 + 9)$

Ⓒ $5(6 + 3)$ Ⓓ $3(10 + 3)$

Ⓔ $15(2 + 3)$

4. Which of the following is equivalent to $8(3 + 2x)$?

Ⓐ $24 + 16x$ Ⓑ $3 + 16x$

Ⓒ $24 + 2x$ Ⓓ $8 + 16x$

Ⓔ $48 + x$

1. Christine saved $180 out of her allowance. She spent $180 for a trip. How much money does she have at that point?

2. If 24 represents the elevation above sea level by 24 feet, what number represents the elevation below sea level by 13 feet?

3. Megan was counting the number of candies she got after trick or treating. Her brother took 3 candies from her candies. What number represents this situation?

4. Emily was standing on a stair. She went 5 steps up and 5 steps down. What number represents the change of her position on a stair?

1. Which of the following is in quadrant II?

Ⓐ $(5, -2)$ Ⓑ $(-1, -4)$

Ⓒ $(-6, 5)$ Ⓓ $(5, 2)$

Ⓔ $(0, 0)$

2. What is the number x on a number line below?

-6 x -5

3. Where is -7.15 located on a number line below?

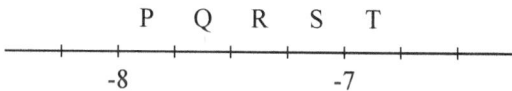

P   Q   R   S   T

-8   -7

4. What are the coordinates of the following point?

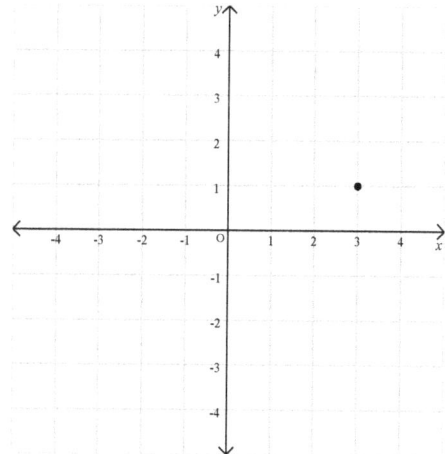

Skill Practice: Understand opposite signs of numbers on the number line (6.NS.C.6.A)

Name: _____

Date: _____

1. Which of the following indicates the opposite of 8 on the number line?

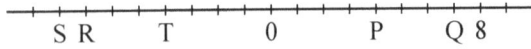

```
+--+--+--+--+--+--+--+--+--+--+--+--+--+--+
   S  R     T        0        P      Q  8
```

2. What is the opposite of the opposite of 39?

3. Which letter indicates the location of the opposite of −8?

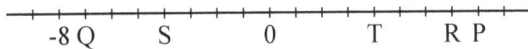

```
+--+--+--+--+--+--+--+--+--+--+--+--+--+--+
  -8  Q     S        0        T      R  P
```

4. What is the opposite of 0?

1. The ordered pair of a point P is $(9, -4)$. What is the ordered pair of a point when P is reflected across x axis?

2. If (a, b) is in quadrant III, in which quadrant is $(-a, b)$?

3. If the sign of x coordinate of a point is positive and the sign of y coordinate of a point is negative, which quadrant does this point locate?

4. The point P is located as below.

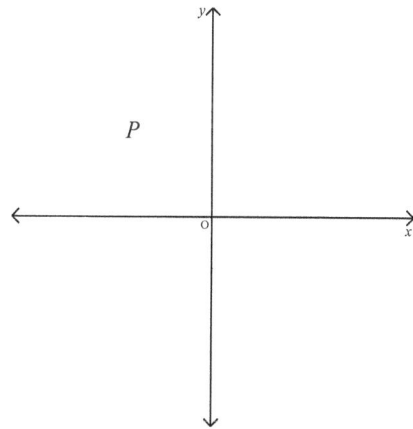

If P is reflected across both axes, which coordinate(s) will change the sign?

1. Which letter represents the correct position of the ordered pair (5, 3)?

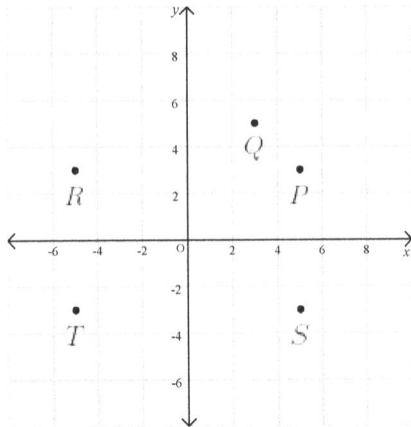

2. Which of the following represents $-4\frac{5}{7}$?

3. What fraction represents the dot on the number line below?

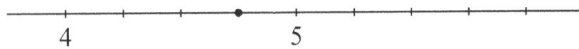

4. What are the coordinates of the point below? Assume that the point is in the center of the box.

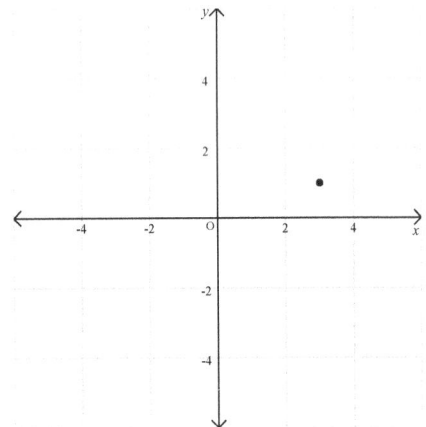

Skill Practice: Understand absolute value of rational numbers
(6.NS.C.7)

Name: _____

Date: _____

---

1. Which of the following is leftmost on the number line?

ⓐ $-|6.1|$ ⓑ 3.5 ⓒ $\dfrac{4}{5}$

ⓓ -3.9 ⓔ $|6|$

2. Which of the following is true?

ⓐ $-4 < -1$ ⓑ $|-4| < 0$

ⓒ $-|-6| > 0$ ⓓ $|-5| < -3$

ⓔ $6 < -8$

3. Which of the following represents the most debt?

ⓐ $-\$133$ ⓑ $-\$39$ ⓒ $-\$98$

ⓓ $-\$145$ ⓔ $-\$33$

4. What is the value of $2 - |-9| - 6$?

Skill Practice: Interpret statements of inequality as the relative position of two numbers on a number line (6.NS.C.7.A)

Name: _____

Date: _____

1. What number is 5 units left of the number -1 ?

2. Which of the following is the correct statement of $-5 < -4$ on a number line?

   Ⓐ -5 is located 4 units to the right

   Ⓑ -5 is located to the left of 4

   Ⓒ -5 is located to the right of -4

   Ⓓ -5 is located 4 units to the left

   Ⓔ -5 is located to the left of -4

3. What expression represents "2 is located to the right of -1 on a number line"?

4. Which of the following inequality represents any number that is located to the right of 2 on the number line? Use x for any number.

   Ⓐ $x < 2$ Ⓑ $x < -2$

   Ⓒ $x > 2$ Ⓓ $x \leq 2$

   Ⓔ $x > -2$

1. Which of the following is the correct statement of the graph on the number line?

Ⓐ $x = 0$ Ⓑ $x < 0$

Ⓒ $x \leq 0$ Ⓓ $x \geq 0$

Ⓔ $x > 0$

2. To take an archery class, you must be 13 years old or older. Joanne is a years old and she is taking an archery class. What expression represents possible age for her?

3. Ian had 26 candies and he ate some of them. If x is the number of candies that he now has, what expression represents the possible values of x?

4. What is the correct expression for "$-5\,^{o}C$ is warmer than $-9\,^{o}C$"?

Ⓐ $9 > 5$ Ⓑ $-9 > -5$

Ⓒ $9 < 5$ Ⓓ $9 = 5$

Ⓔ $-9 < -5$

Skill Practice: Understand the absolute value of a rational number as its distance from 0 on the number line (6.NS.C.7.C)

Name: _____

Date: _____

1. What is the value of $2 - |-12|$?

2. Joan's house is 4.1 miles East of the library, while Michael's house is 5.7 miles West of the library. What is the distance between Joan's house and Michael's house?

3. What is the value of $-|-3.4|$?

4. Jim and Bob played a game on a number line. They both started at the origin. Jim moved his marble 8 units to west and Bob moved his marble 5 units to east. Which expression represents the distance between two marbles?

  Ⓐ $8 - 5$ Ⓑ $|-8| + |5|$

  Ⓒ $-8 + 5$ Ⓓ $|-8 + 5|$

  Ⓔ $|8 - 5|$

1. The following table shows bank balances for customers. Who has a debt greater than $3740?

Name	Balance in dollars
Jane	-3770
Rachel	-3650
Michael	-3690
Ian	-3720

2. Winter of this year was colder than winter of last year. The last year's average temperature in Celsius was -16 degree. If t is the average temperature for winter of this year, which expression represents this situation?

3. Tim dived in the ocean 5 times. He recorded all the 5 numbers in feet that he read on an elevation dial. Which of the following numbers represent the deepest dive?

Ⓐ -40 Ⓑ -32 Ⓒ -33

Ⓓ -50 Ⓔ -66

4. Thomas checked his bank account. The total balance on his bank account last month was $-\$90$. This month, his spending was greater than his income. Which of the following could be the total balance of his account?

Ⓐ $\$175$ Ⓑ $\$0$ Ⓒ $-\$26$

Ⓓ $\$70$ Ⓔ $-\$233$

1. What is the sum of the distances between P and Q, Q and R, and R and S?

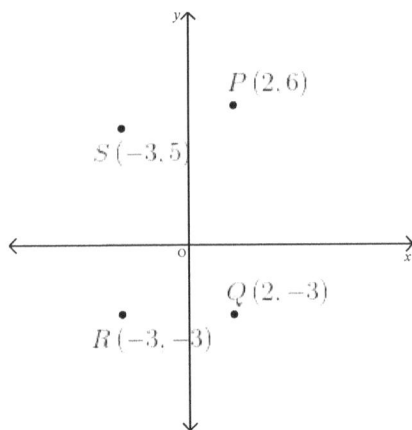

$P\,(2,6)$

$S\,(-3,5)$

o

$Q\,(2,-3)$

$R\,(-3,-3)$

2. Megan plotted the position for herself as $(0,\ 0)$. Then, she found the her house is located at $(-4,\ 7)$ and the school is located at $(3,\ 7)$. What is the distance between her house and the school?

3. Katie plotted the position for herself as $(0,\ 0)$. Then, she found that her house is located at $(-3,\ 5)$, Tim's house is located at $(9,\ 5)$, and the school is located at $(9,\ -3)$. What is the distance between her house and the school if she dropped by Tim's house?

4. What is the distance between $(3,\ -5)$ and the y axis?

Name: _____

Date: _____

1. Fred walked $\dfrac{5}{9}$ of the way from his home to park. If he walked $\dfrac{1}{8}$ mile, what is the distance between his home and park?

Ⓐ $\dfrac{49}{9}$ Ⓑ $\dfrac{40}{9}$ Ⓒ $\dfrac{7}{40}$

Ⓓ $\dfrac{11}{40}$ Ⓔ $\dfrac{9}{40}$

2. Bob and 3 friends went on a trip. Each of them paid $46 for expense. They spent $41 for gas and $52.4 for food. When the trip was over, they splitted the remaining amount. How much did each of them get?

Ⓐ $30.2   Ⓑ $45.53   Ⓒ $68

Ⓓ $22.65   Ⓔ $14.87

3. Which of the following is equivalent to $60 + 10x$?

Ⓐ $10(6 + x)$ Ⓑ $5(6 + x)$

Ⓒ $10(6 + 10x)$ Ⓓ $5(6 + 10x)$

Ⓔ $2(6 + x)$

4. Ian has invested in the stock market. He gained $35 at first month, but lost $72 at next month. What number represents the total amount of loss or gain for two months?

Ⓐ $-$107$ Ⓑ $-$37$ Ⓒ $37

Ⓓ $0   Ⓔ $107

5. What is the value of the expression,
$7 + (-7) - (-6)$?

Ⓐ 20 Ⓑ -6 Ⓒ 6

Ⓓ 8 Ⓔ 7

6. What are the coordinates of an ordered pair B when the ordered pair A, $(-4, -7)$ is reflected across y axis?

Ⓐ $(7, -4)$ Ⓑ $(4, 7)$

Ⓒ $(4, -7)$ Ⓓ $(-4, 7)$

Ⓔ $(-7, -4)$

7. The maximum weight allowed inside an elevator is 940 lb. Which of the following represents this situation if w is the total weight(lb) of the people inside an elevator?

Ⓐ $w > 940$ Ⓑ $w = 940$

Ⓒ $w \geq 940$ Ⓓ $w \leq 940$

Ⓔ $w < 940$

8. Where is the ordered pair $(-3.7, -3.3)$ located?

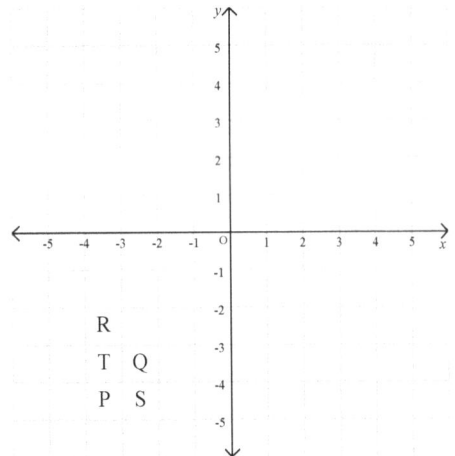

Ⓐ P Ⓑ R Ⓒ S

Ⓓ Q Ⓔ T

9. Ian has 6.84 lbs of salt. If he poured the same amount in 4 containers, how many pounds of salt is in each container?

Ⓐ 1.71   Ⓑ 2.21   Ⓒ 1.21

Ⓓ 0.171   Ⓔ 1.31

10. Ave's restaurant used $1\frac{5}{7}$ lbs of flour to make one extra large pizza. If he has $6\frac{3}{4}$ lbs of flour, how many whole extra large pizza can he make?

Ⓐ 3   Ⓑ 4   Ⓒ 5

Ⓓ 2   Ⓔ 8

11. Find the greatest common factor of 30 and 18.

Ⓐ 3   Ⓑ 6   Ⓒ 18

Ⓓ 2   Ⓔ 1

12. If $(a, -b)$ is in quadrant II, where is $(-a, b)$?

Ⓐ Quadrant III   Ⓑ Quadrant I

Ⓒ Can not determine   Ⓓ Quadrant II

Ⓔ Quadrant IV

13. Joanne is making cookie baskets and wants to make all the cookie baskets identical without having any pieces of cookie left over. If Joanne has 24 chocolate cookies and 42 sugar cookies, what is the greatest number of cookie baskets Joanne can make?

Ⓐ 6　　　Ⓑ 24　　　Ⓒ 42

Ⓓ 4　　　Ⓔ 7

14. The coordinates of Ethan's position were (6, 6). If he walked 7 units south and 9 units west, what are the coordinates of his position now?

Ⓐ $(-3, 13)$　　　Ⓑ $(-1, -3)$

Ⓒ $(15, -1)$　　　Ⓓ $(-3, -1)$

Ⓔ $(-1, 15)$

15. Andrew bought 19 shirts with $14.59 each and 17 pants with $20.79 each for the online store. How much did he spend in all?

Ⓐ $640.64　　　Ⓑ $620.64　　　Ⓒ $630.64

Ⓓ $645.64　　　Ⓔ $650.64

16. The average temperature in summer is 35^oC. If the average temperature in winter is 38^oC lower than the average temperature in summer, what is the average temperature in winter?

Ⓐ 4^oC　　　Ⓑ 0^oC　　　Ⓒ 3^oC

Ⓓ -4^oC　　　Ⓔ -3^oC

Expressions and Equations

6.EE

Skill Practice: Write and evaluate numerical expressions with whole-number exponents (6.EE.A.1)

Name: _____

Date: _____

1. What is equivalent expression using whole number exponents for the expression, $5 \times 5 \times 5 + 3 \times 3 \times 3 + 6 \times 6$?

2. What is the value of $5^2 + 4^0$?

3. Evaluate $\left(\dfrac{4}{5}\right)^2$ and write an answer as a fraction.

4. Evaluate $6 - 3 \times 8 + (7 - 6) + 3^2$.

1. The area of a triangle with a base b and a height h is $A = \dfrac{1}{2}bh$. What square units is the area of a triangle with a base 14 units and a height 6 units?

2. Celsius (^{o}C) and Fahrenheit (^{o}F) can be converted using the formula $C = \dfrac{5}{9}(F - 32)$. What is Celsius degree if Fahrenheit degree is 56? Round to the nearest whole number if necessary.

3. What is the value of $x(x - y) + y^2$ when $x = 7$ and $y = 3$?

4. Evaluate the expression "6 is subtracted from the product of 6 and x" when $x = 8$.

1. What is the expression for " 12 less than twice of *n* " ?

2. Kristy gave away *b* buttons out of 35 buttons she had. How many buttons does Kristy have now?

3. What is the expression of the product of 6 and the sum of *x* and 9?

4. Write the expression for "15 more than the product of *y* and 3".

---

1. What is the coefficient of z in the expression $-2z$?

2. How many terms are in the expression
$9 - 3h + 4h^4$?

---

3. What is the coefficient of m in the expression
$2n - 9m$

4. What is the expression of " product of factors, t,
$5 + h$ "?

1. If $u = 12$, what is the value of $2u - 13$?

2. The volume of a cube with side lengths of x is $V = x^3$. What is the volume of a cube with side lengths of $\frac{1}{4}$?

3. Find the value of an expression $x + v - 5.9$ for $x = 9.3$ and $v = 5.1$.

4. Evaluate the expression $x^4(x - z)$ when $x = -1$ and $z = 5$.

1. Simplify $(6y + 7) - (6y - 2)$.

2. Simplify the expression $x + 6x + 7y + y$.

3. Which of the following is equivalent to $54a + 12b$?

   Ⓐ $6(9a + b)$ Ⓑ $6ab$

   Ⓒ $6(a + b)$ Ⓓ $6(9a + 2b)$

   Ⓔ $6(a + 2b)$

4. Simplify the expression $3(9x + 7)$ using the distributive property.

1. Simplify the expression $6(9x - 3y) - 7x + 3y$.

2. Simplify $y \times y + y \times y \times y + y + 3y$ using exponents.

3. What is the simplified expression of $5(3x) - 2x + x + 8$?

4. Simplify the expression
$5y \times y + 4y \times y \times y + y + 6y$ using exponents.

1. Choose the equation or inequality where $x = 8$ is a solution.

Ⓐ $26 > 3 + 5x$ Ⓑ $2x - 4 = 12$

Ⓒ $4x + 4 \leq 9$ Ⓓ $2 - (x + 3) = -3$

Ⓔ $4x + 7 = 11$

2. What is the solution to the equation $4x - 14 = 14$?

3. What is the value of x that satisfies the equation $3 - (x + 1) = 5x - 28$?

4. Ave has $15 to spend at the Fair. He wants to ride a roller coaster which costs $5.89. What is the maximum number of times he can ride on a roller coaster?

1. If Alan can finish painting a room in h hours, what is the expression of the amount of job he can finish in one hour?

2. Emily drove a car with 45 mph. What is the expression of the total distance she drove if she drove h hours?

3. If a boat travels 15 mph in still water and the speed of the current is c mph, what is the real speed of the boat when the boat is traveling downstream?

4. If Ethan can finish painting his room in n hours and John can paint the same room twice as fast as Ethan, how long would it take John to finish painting the room?

1. Ethan went to a book store and his mother gave him $8. Then he bought a book that costs $10.74 and had $27.26 left. How much did he have before his mother gave him $8?

2. Kristy bought 7 drinks that each cost the same amount. If she spent $21, how much did she spend on each drinks? (Write and solve an equation to solve the problem.)

3. Gregory wanted to buy one book and one toy. He found that a toy costs $5.19 more than a book. If a toy costs $10, how much does a book cost?

4. Joanne's father is 5 times as old as her. In 3 years, he will be 4 times as old as her. How old is Joanne now?

1. Morgan owns a store which sells accessories. She wants to have at least $2860 as profit per month. If p represents a profit, what is the correct expression of this situation using an inequality?

2. Jenny counted how much she saved and it was more than $23. Which of the following represents the amount she saved?

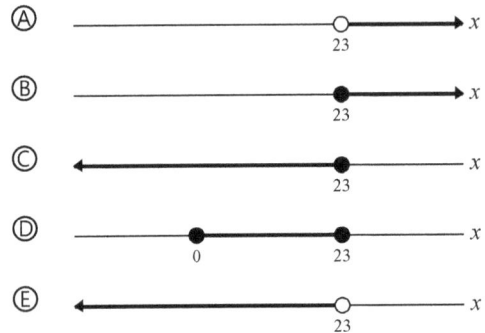

3. Christine counted the number of books she has and it was more than 100. If n is the number of books that Christine has, what is the expression of this situation?

4. Dorothy's family saved money for a trip to Disney World. When they counted how much they saved, they figured that they could spend $840 for hotel. They planed to stay seven nights. Assuming that each night costs the same, which inequality shows the price of one night hotel that they can stay? Let x be the one night hotel price.

1. Each chocolate bar costs $4 and each box contains 5 chocolate bars. How many boxes of chocolate bars can be purchased with $260?

2. The price of one pen is $3.89. If y is the price for x number of pens, what is the equation between x and y?

3. Matthew walked to the park with a rate of 2 mph. Which of the following graph shows the relation between the distance he walked and the time?

Ⓐ
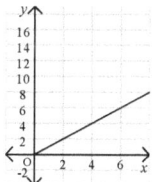

Ⓑ
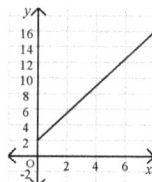

Ⓒ
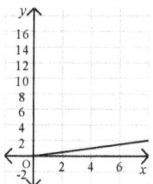

Ⓓ
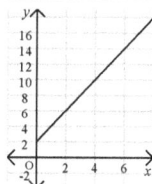

Ⓔ
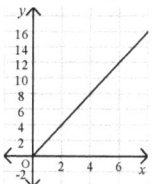

4. What is the equation for the following data in the table?

n	C
1	4
2	8
3	12
4	16

1. What is the value of $a^3 + b^5$ when a equals 4 and b equals 3 ?

   Ⓐ 180   Ⓑ 206   Ⓒ 56

   Ⓓ 27   Ⓔ 307

2. The weight of one can is 12 ounces. If y represents the weight of x cans in ounces, what is the equation between x and y?

   Ⓐ $y = x + 12$ Ⓑ $y = x - 12$

   Ⓒ $y = \dfrac{1}{12} \times x$ Ⓓ $y = 12 \times x$

   Ⓔ $x = 12 \times y$

3. What is the expression of "the sum of 17 and the twice of x is 8 less than y"?

   Ⓐ $17 + 2x = y - 8$ Ⓑ $17 + 2x = 8 - y$

   Ⓒ $2(17 + x) = 8 - y$ Ⓓ $17 + x + 2 = y - 8$

   Ⓔ $2(17 + x) = y - 8$

4. Which of the following x value does *not* satisfy the inequality, $6 - 2x < 2 + 4x < 20 - 2x$?

   Ⓐ 4   Ⓑ 1.1   Ⓒ 1.9

   Ⓓ 1.5   Ⓔ 2.8

5. What is the equation of the following graph?

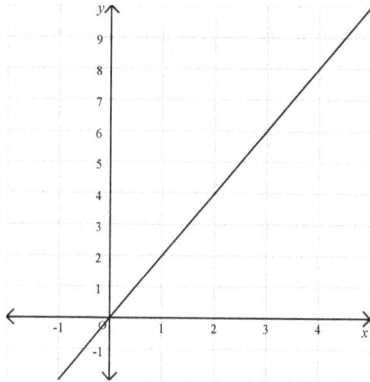

Ⓐ $y = \dfrac{x}{2}$

Ⓑ $y = x + 2$

Ⓒ $y = 2x$

Ⓓ $y = x - 2$

Ⓔ $y = \dfrac{2}{x}$

6. Ethan wanted to buy one book and one toy. He found that a toy costs $4 more than a book. If a book costs $19, how much did he spend total?

Ⓐ $37.95   Ⓑ $34   Ⓒ $42

Ⓓ $38   Ⓔ $46.05

7. Paul went to an amusement park and spend $50 for the ticket. Also, he spent $5.54 for lunch and $15 for parking. If x is the amount he had before he went to an amusement park, what inequality represents the amount x?

Ⓐ $x \geq 59.46$ Ⓑ $x < 70.54$

Ⓒ $x \leq 59.46$ Ⓓ $x \leq 70.54$

Ⓔ $x \geq 70.54$

8. Megan's allowance each week is $6. She saved all of her allowance to buy a camera. What is the total amount she saved for n weeks? Use P as the total amount.

Ⓐ $p = 6n + 6$ Ⓑ $p = n + 12$

Ⓒ $p = \dfrac{n}{6}$ Ⓓ $p = n + 6$

Ⓔ $p = 6n$

9. What is the value of the expression,

$6 - 5 \times a + a^2 - 2a$, if $a = 3$?

Ⓐ -6 Ⓑ -3 Ⓒ 6

Ⓓ -9 Ⓔ 3

10. If $y = 4$, what is the value of $20 + 13y$?

Ⓐ 72 Ⓑ 52 Ⓒ 92

Ⓓ 37 Ⓔ 69

11. Simplify the expression $4(5x) - 4(x - 1) + 1$.

Ⓐ $24x + 5$ Ⓑ $16x + 5$

Ⓒ $24x - 3$ Ⓓ $16x - 5$

Ⓔ $16x - 3$

12. If a boat travels 15 mph in still water and the speed of the current is c mph, what is the real speed of the boat when the boat is traveling upstream?

Ⓐ $15c$ Ⓑ $15 + c$ Ⓒ $\dfrac{c}{15}$

Ⓓ $15 - c$ Ⓔ $\dfrac{15}{c}$

13. In a store, there are two sizes of water bottles. The smaller water bottle weighs 8oz and the larger water bottle weighs 14oz. If one cart can hold 20 lb and 20 large water bottles are already on the cart, what is the maximum number of small water bottles that can be put on the cart?

Ⓐ 5   Ⓑ 7   Ⓒ 8

Ⓓ 6   Ⓔ 4

14. Simplify the expression $(14x + 9) - 2(7x - 6)$.

Ⓐ 3   Ⓑ −3   Ⓒ 21

Ⓓ −12   Ⓔ 15

15. After the trick or treating, Andrew and Katie counted the number of candies they got. They found that Katie had 132 candies which is twice as many candies as Andrew had. How many candies did Andrew have? Use an equation to find the answer.

Ⓐ 130   Ⓑ 264

Ⓒ 66   Ⓓ 134

Ⓔ 33

16. Gregory wanted to buy a camcorder that costs $100. If x is the amount he has to save, which inequality represents enough money to buy camcorder?

Ⓐ $x \le 100$ Ⓑ $x = 100$

Ⓒ $x \ge 100$ Ⓓ $x < 100$

Ⓔ $x > 100$

Geometry

6.G

1. Joanne has a rectangular garden with a width of 18 feet and the length of 6 feet. In one day, she put a fence as shown below to plant flowers in one triangular garden and vegetable in the other triangular garden. What is the area of the garden with flowers in square feet?

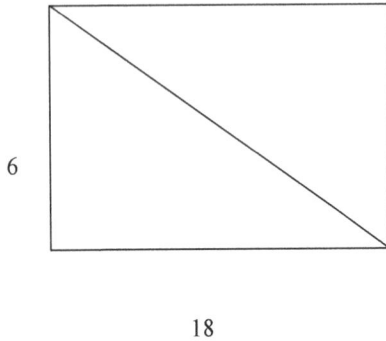

6

18

2. What square unit is the area of the shaded triangle below?

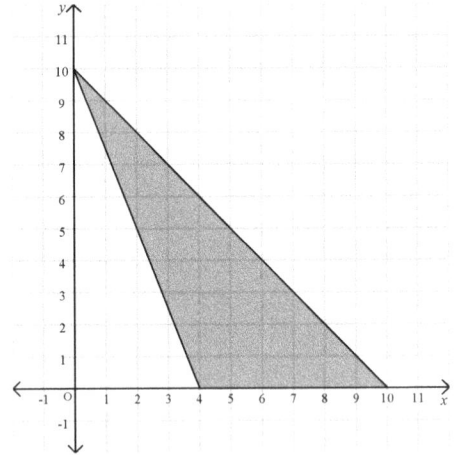

3. Find the area of the following trapezoid.

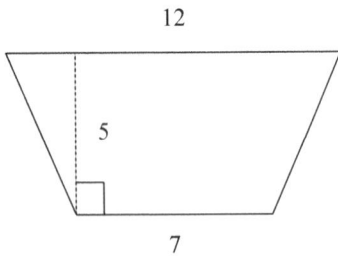

12

5

7

4. Ave cuts grass at $1 per square yard. How much would he earn cutting the grass for the area below? All the lengths given below are in yards.

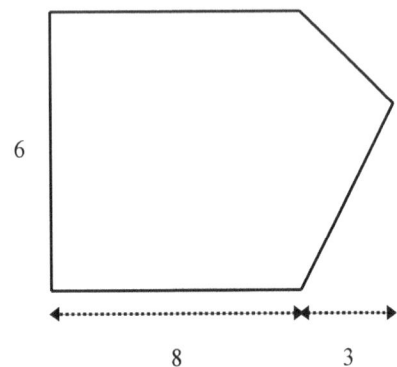

6

8   3

1. What is the volume of a box in cubic feet if a box has a height of 8 feet, a width of 3 feet, and a length of 2 feet?

2. Ave wanted to pour water in a cube whose side is 5 inches. How much of cubic inches of water can he pour in this cube?

3. The size of a pool is $8\frac{1}{2}$ feet wide, 5 feet long, and $4\frac{1}{2}$ feet high. The pool is drained and refilled every night. What cubic feet volume of water would be required to fill the pool to the rim?

4. A truck has a rectangular prism shape of cargo in the back. The dimensions of the cargo are 10.5 feet by 17.5 feet by 3.5 feet. Cube shaped boxes with 3.5 feet will be loaded into the cargo. How many boxes will completely fill the cargo?

1. If three vertices of a rectangle are $(-3, 3)$, $(6, 3)$, and $(6, -6)$, what are the coordinates of the fourth vertex?

2. If three vertices of a rectangle are $(-3, 6)$, $(-3, -8)$, and $(6, -8)$, what is the area of this rectangle in square units?

3. Fred was drawing a map for his town. He found that his house, his school, the library, and the post office make a rectangle. So, he plotted his house at $(-3, 7)$ and the library at $(3, -8)$, and the sides of a rectangle in the coordinate plane are parallel to the axes. what is the perimeter of this rectangle?

4. Three vertices of a trapezoid are $(-2, 4)$, $(3, 4)$, and $(-6, -4)$. Which of the following coordinates are possible to be the fourth vertex of a trapezoid?

Ⓐ $(5, -4)$ Ⓑ $(5, -3)$

Ⓒ $(-4, 5)$ Ⓓ $(4, -3)$

Ⓔ $(4, -2)$

1. Which of the following does *not* include a rectangle in its net?

Ⓐ Cube   Ⓑ Triangular prism

Ⓒ Square pyramid   Ⓓ Rectangular prism

Ⓔ Cone

2. What is the surface area of the following square pyramid, if the side of a square is 8 and the height of a triangle is 4?

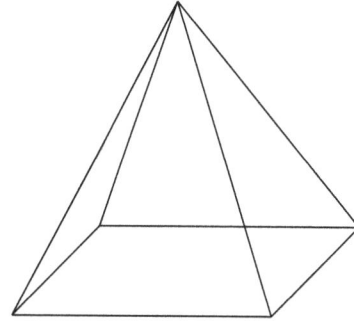

Figure not drawn to scale

3. Christine wanted to wrap a gift box whose dimension is 12 inches for the length, 4 inches for the width, and 4 inches for the height. What is the surface area in square inches of this gift box?

4. What is the surface area of a following triangular prism?

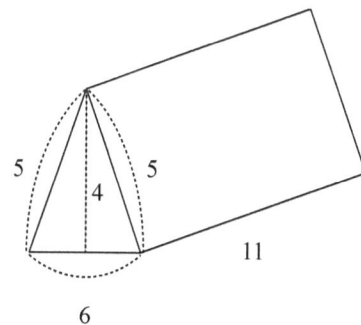

5   4   5

11

6

Figure not drawn to scale

1. What square unit is the area of the parallelogram below?

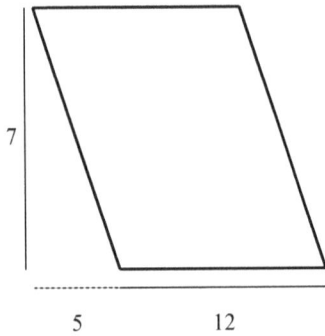

Ⓐ 70　　　Ⓑ 119　　　Ⓒ 35

Ⓓ 168　　　Ⓔ 84

2. A rectangular prism consists of 32 cubes. Each cube has a side length of 3cm. What is the volume of a rectangular prism in cm^3?

Ⓐ 1728cm³　　　Ⓑ 96cm³　　　Ⓒ 54cm³

Ⓓ 864cm³　　　Ⓔ 432cm³

3. Jimmy is filling a rectangular box with sugar cubes. If the dimensions of the box are 15" by 30" by 10" and the sugar cubes have side length of 5", then how many sugar cubes can fit in the box?

Ⓐ 18　　　Ⓑ 36　　　Ⓒ 900

Ⓓ 90　　　Ⓔ 180

4. What is the area of the following polygon which consists with a right triangle and a rectangle?

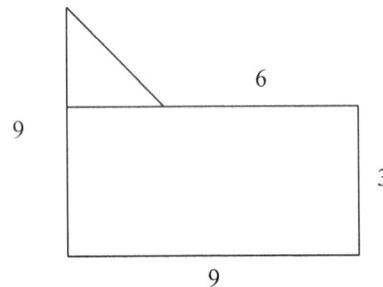

Figure not drawn to scale

Ⓐ 22.5　　　Ⓑ 36　　　Ⓒ 45

Ⓓ 67.5　　　Ⓔ 40.5

5. Bob and his friends are building a scale model in order to plan out the project. The dimension of model building are $2\frac{1}{2}$ inches high, and $3\frac{1}{2}$ inches wide, and $1\frac{1}{2}$ inches long. What cubic inches is the volume of a model building?

Ⓐ $\dfrac{109}{8}$ Ⓑ $\dfrac{107}{8}$ Ⓒ $\dfrac{105}{8}$

Ⓓ $\dfrac{117}{8}$ Ⓔ $\dfrac{119}{8}$

6. Megan found that her house, her school, the library, and the fire station make a parallelogram on a map. If she plotted her house as $(-3,\ 5)$, her school as $(9,\ 5)$, and the library as $(-8,\ -4)$, what are the coordinates of the fire station?

Ⓐ $(4,\ 4)$ Ⓑ $(-4,\ 4)$

Ⓒ $(4,\ -4)$ Ⓓ $(9,\ -4)$

Ⓔ $(-4,\ 9)$

7. Paul is making a rectangular prism stand with the length 3 feet, the width 2 feet, and the height 3 feet. He wanted to paint all sides except the base. If the paint costs $0.51 for each square feet, how much would it cost for Paul to paint the stand?

Ⓐ $10.71   Ⓑ $24.48

Ⓒ $13.77   Ⓓ $21.42

Ⓔ $18.36

8. What is the sum of all the edges, in cm, when this net is folded up?

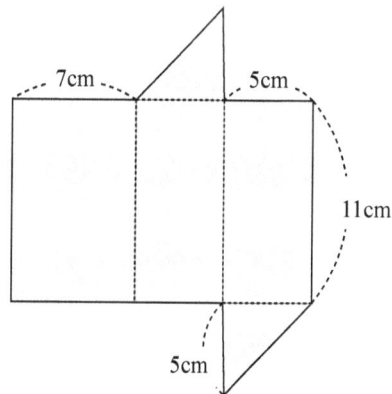

Ⓐ 62cm   Ⓑ 78cm   Ⓒ 28cm

Ⓓ 56cm   Ⓔ 67cm

Statistics and Probability

6.SP

Skill Practice: Recognize statistical questions (6.SP.A.1)

Name: _____
Date: _____

1. Which of the following is the most appropriate as a statistical question?

   Ⓐ How tall are the students in my class?

   Ⓑ What is my height?

   Ⓒ What is my salary?

   Ⓓ What is my final grade?

   Ⓔ How much is an apple?

2. Ethan's teacher has finished grading her students' exams. She said students could ask statistical questions. Which of the following is a statistical question?

   Ⓐ How many students did not complete the exam?

   Ⓑ How many students earned a C?

   Ⓒ What was the lowest score?

   Ⓓ How did this class do compare to other classes?

   Ⓔ How were the scores on the multiple choice questions?

3. Which of the following is a non-statistical question?

   Ⓐ How many books did every student read over summer break?

   Ⓑ How many miles do your coworkers drive to work?

   Ⓒ What were the students Math test scores?

   Ⓓ How many miles did Megan walk yesterday?

   Ⓔ How many hours did your family members sleep last night?

4. Your school principal comes to visit your class, and you want to take the opportunity to ask him some statistical questions. Which of the following is a statistical question that you could ask?

   Ⓐ How many hours does each student spend in school?

   Ⓑ How many teachers were hired last year?

   Ⓒ How many students registered this year?

   Ⓓ How many students graduated last year?

   Ⓔ What is the statewide standard math test score for 4th grade?

Skill Practice: Understand a set of data has a distribution described by its center, spread, and overall shape (6.SP.A.2)

Name: _____

Date: _____

1. The following figure shows the number of students with test scores. Based on the graph, what score occurs the most?

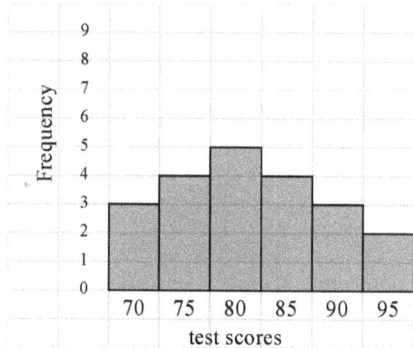

2. Emily has 5 children. When she measured the height of 5 children, she got 32, 50, 55, 60, 64 inches. What is the center of this data?

3. Which of the following is the _least_ appropriate characteristic of a data set that a statistical question would like to answer?

Ⓐ Overall shape of a distribution

Ⓑ Center of a distribution

Ⓒ Variability of a distribution

Ⓓ Spread of a distribution

Ⓔ Area of a distribution

4. The following figure shows the number of students with test scores. Based on the graph, which of the following is correct?

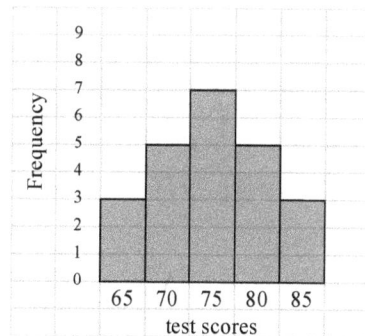

Ⓐ The center of this data is 65

Ⓑ The test score that occurs the most is 85

Ⓒ The data does not spread at all

Ⓓ Overall shape of the data is symmetric

Ⓔ There are 26 students

1. Which of the following is the most appropriate measure which describes how its value varies with a single number?

   Ⓐ A measure of shape

   Ⓑ A measure of center

   Ⓒ None of other choices

   Ⓓ A measure of cost

   Ⓔ A measure of variation

2. The following figure shows the number of students with test scores. Based on the graph, how much do the data vary from the center to the greatest value?

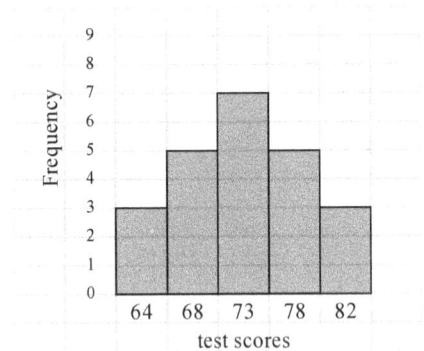

3. Which of the following data has the greatest variation?

   Ⓐ {30, 32, 34, 36, 38}

   Ⓑ {39, 40, 41, 42, 43}

   Ⓒ {24, 27, 30, 33, 36}

   Ⓓ {17, 21, 25, 29, 33}

   Ⓔ {17, 22, 27, 32, 37}

4. Which of the following value summarizes all of its values with a single number?

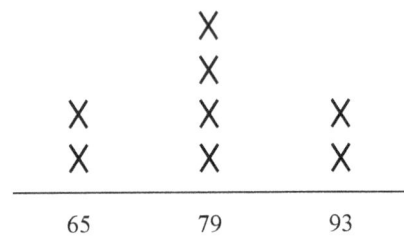

   Ⓐ 79   Ⓑ 93   Ⓒ 65

   Ⓓ 28   Ⓔ 14

1. Which of the following is the most appropriate name of the given figure?

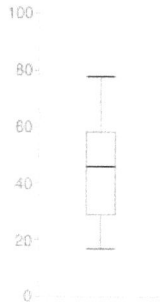

Ⓐ Box plot   Ⓑ Pie chart

Ⓒ Histogram   Ⓓ Dot plot

Ⓔ Line chart

2. Which type of graph would best show the number of books borrowed from the library each month for first quarter?

Ⓐ Circle graph   Ⓑ Box plot

Ⓒ Line graph   Ⓓ Bar graph

Ⓔ Stem and Leaf plot

3. The box plot for a data is given below.

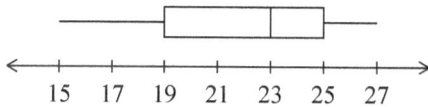

15 17 19 21 23 25 27

What is the median for the data?

4. Which type of graph would best show the number of students who like either vanilla ice cream, chocolate ice cream, or strawberry ice cream?

Ⓐ Bar graph   Ⓑ Circle graph

Ⓒ Box plot   Ⓓ Line graph

Ⓔ Stem − and − leaf

1. Kristy recorded all of her math test scores through out the year. What is the median of her test scores?

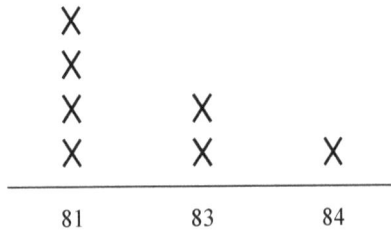

```
 X
 X
 X       X
 X       X       X
_____
 81      83      84
```

2. Larry recorded the number of jump rope he did for 5 days using a line graph below. What is the mean number of jump rope he did every day?

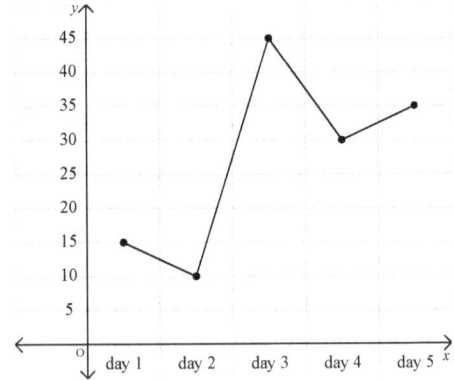

3. Tom recorded the number of jump rope he did for 5 days using a line graph below. What is the range of the number of jump rope he did?

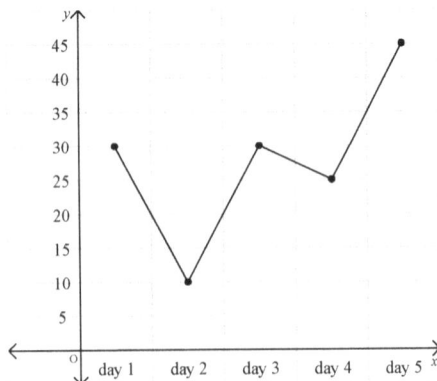

4. Mrs. Rachel wants to analyze heights of students in her classroom. Which of the following is the _least_ appropriate way of summarizing the data set?

Ⓐ Reporting interquartile range

Ⓑ Reporting median

Ⓒ Reporting deviation scores

Ⓓ Reporting mean

Ⓔ Reporting standard deviation

Name: _____

Date: _____

1. The following figure shows the number of students with test scores. Based on the graph, how many students took the test?

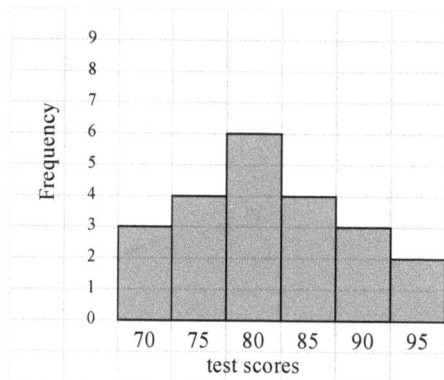

Frequency
test scores

2. Violet recorded all of her math test scores through out the year. How many did she take over the year?

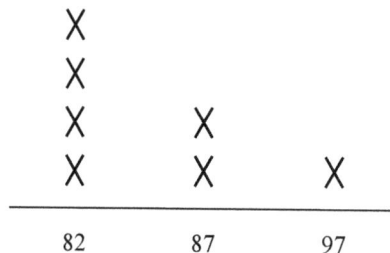

X
X
X   X
X   X   X

82   87   97

3. Fred recorded the number of jump rope he did for 5 days using a line graph. How many jump rope did he do for 5 days?

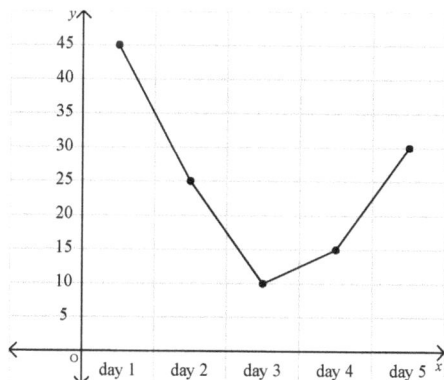

day 1   day 2   day 3   day 4   day 5

4. The following figure shows the number of students with test scores. Based on the graph, how many students got 90 points or higher?

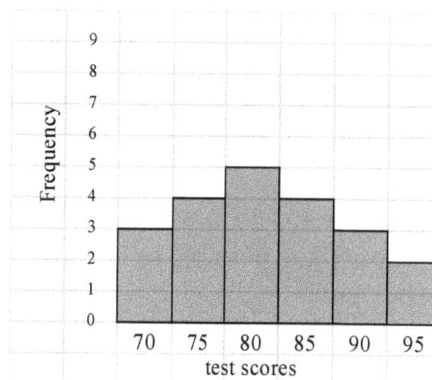

Frequency
test scores

1. The given figure summarizes students' blood type in a school. Which of the following is the best description of the number for blood type A, 90?

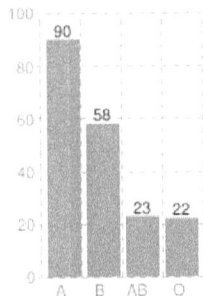

Ⓐ There are 90 students with the blood type A

Ⓑ There are 90 blood types

Ⓒ Mode of the blood type A is 90

Ⓓ Median of the blood type A is 90

Ⓔ Mean of the blood type A is 90

2. Larry recorded the number of jump rope he did for 5 days using a line graph below. What day did he do the least amount of jump rope?

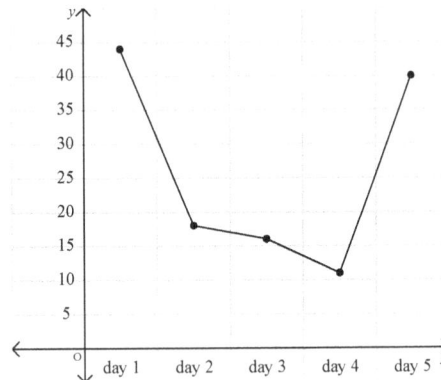

3. Ms. Megan asked her 6th grade students the number of books they read over the summer break, and displayed the data using a box plot below.

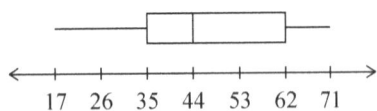

If Julie was the top 15% who read the most, which of the following is possible to be the number of books she read?

Ⓐ 44   Ⓑ 38   Ⓒ 52

Ⓓ 65   Ⓔ 35

4. Dorothy gathered a data about heights in inches and weights in pounds of his students, and made a chart as below.

	heights	weights
Dorothy	53	56
Philip	55	92
Christine	60	77
John	67	99
Jeff	68	111

Which of the following is a correct statement?

Ⓐ Philip is 36 ounces heavier than Dorothy

Ⓑ Jeff is 15 inches taller than Dorothy

Ⓒ The unique mode of this data is 60 inches

Ⓓ John weighs 67 pounds

Ⓔ Christine is 7 inches shorter than Jeff

1. Students' grades data in teacher Mrs. Rachel's classroom is {85, 88, 74, 79, 84}. What is the median of this data set?

2. The following figure shows the number of students with test scores. Based on the graph, what is the mode?

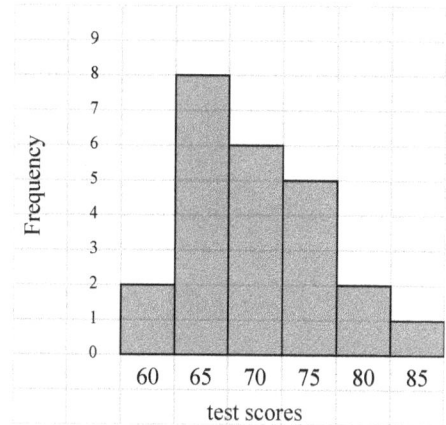

3. Ms. Rachel did a survey about how many minutes her students spent watching TV daily, and then she drew a line plot. What is the outlier of the data?

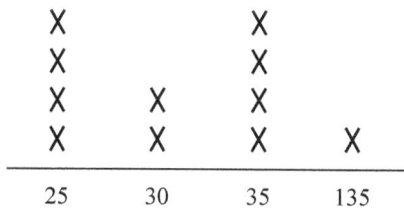

4. The following figure shows the number of students with test scores. Based on the graph, what is the mean of the data?

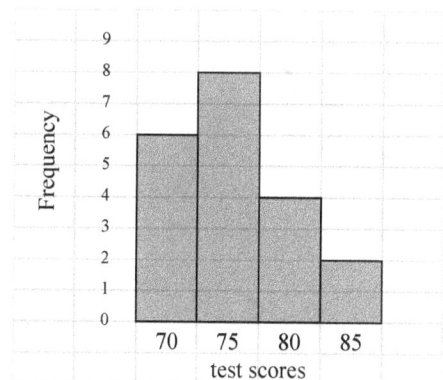

Skill Practice: Relate the choice of measures of center and variability to the shape of the data distribution (6.SP.B.5.D)

Name: _____

Date: _____

1. The given figure is a distribution of the numbers of books read by students during summer break. Which of the following is the *least* relevant explanation of this distribution?

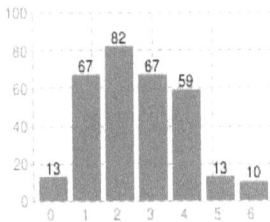

Ⓐ The mode is 2, and the frequency is 82

Ⓑ The median is greater than the mean

Ⓒ This distribution is asymmetric (none symmetric)

Ⓓ The mean is greater than the mode

Ⓔ This distribution is positively skewed

2. The given figure is a distribution of the numbers of books read by students during winter break. Which of the following is the most relevant explanation of this distribution?

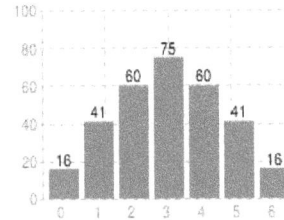

Ⓐ This distribution is asymmetric (none symmetric)

Ⓑ This distribution is positively skewed

Ⓒ The median is greater than the mean

Ⓓ The mean, mode, and median are all same

Ⓔ The mean is greater than the mode

3. Class A and Class B took the same exam. The graphs below show distributions of exam scores. Which of the following statements is correct?

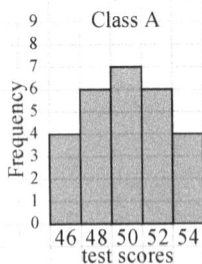

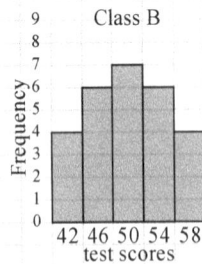

Ⓐ The mean of exam scores for class A is higher than class B

Ⓑ The mode of class A is lower than class B

Ⓒ The range of exam scores for class A is higher than class B

Ⓓ The scores of class A vary less than class B

Ⓔ The class A has more students

4. Class A and Class B took the same exam. The graphs below show distributions of exam scores. Which of the following statements is *not* correct?

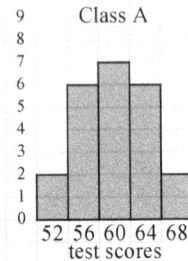

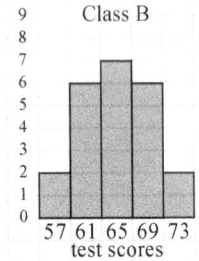

Ⓐ Classes A and B have the same range of exam scores

Ⓑ Class A has a lower mode than class B

Ⓒ Class B has a higher mean than class A

Ⓓ Classes A and B have the same number of students

Ⓔ The scores of class A vary more than the scores of class B

1. Your teacher is giving a talk on how students worked in Math class. Which of the following is statistical question that you could ask your teacher?

Ⓐ How many students finished their works in class?

Ⓑ How many minutes did it take students to finish practice problems?

Ⓒ Which topic did students spend the most time?

Ⓓ What was the average score on the first test?

Ⓔ Who earned the most score?

2. The following figure shows the number of students with test scores. Based on the graph, what score occurs the least?

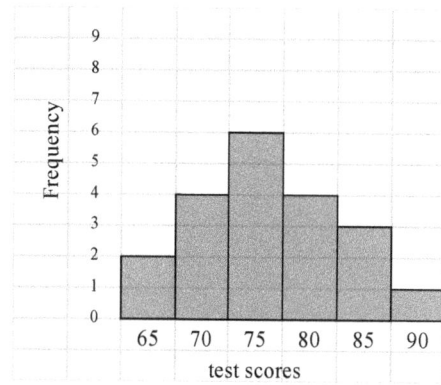

Ⓐ 80   Ⓑ 85   Ⓒ 65

Ⓓ 90   Ⓔ 70

3. What is the difference between the center and the lowest data value?

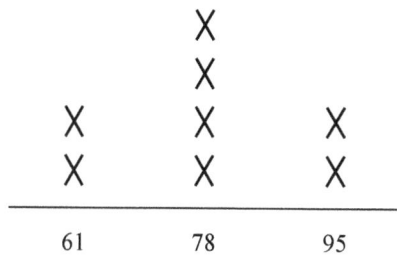

Ⓐ 78   Ⓑ 61   Ⓒ 17

Ⓓ 34   Ⓔ 95

4. Megan recorded all of her math test scores through out the year. What is the range of this data?

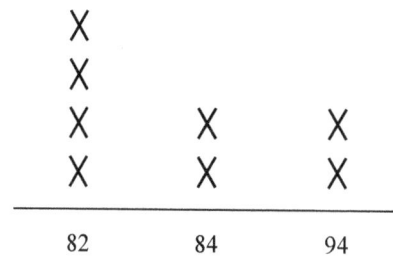

Ⓐ 2   Ⓑ 12   Ⓒ 13

Ⓓ 1   Ⓔ 10

5. The given figure shows the number of a new toy sold in a toy shop over last five months. Based on the graph, which of the following is the the best estimate of the total number of the toy sold during five months?

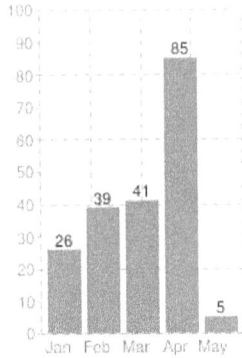

Ⓐ 39.2   Ⓑ 65.2   Ⓒ 196

Ⓓ 104.2   Ⓔ 145.2

6. Students' grades data in teacher Mrs. Rachel's classroom is {83, 84, 73, 76, 78}. What is the mean of this data set?

Ⓐ 80.8   Ⓑ 78.8   Ⓒ 73

Ⓓ 394   Ⓔ 79.8

7. The data for a box plot is given below.

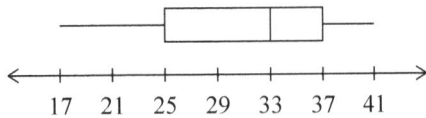

What is the interquartile range for the data?

Ⓐ 16   Ⓑ 8   Ⓒ 4

Ⓓ 24   Ⓔ 12

8. Which of the following is the most appropriate name of the given figure?

Ⓐ Box plot   Ⓑ Dot plot

Ⓒ Line chart   Ⓓ Histogram

Ⓔ Pie chart

9. Which of the following is a statistical question?

Ⓐ How many hours did students in your class sleep last night?

Ⓑ How many miles did Emily drive to home yesterday?

Ⓒ How many emails did Ian send yesterday?

Ⓓ How fast can Matthew run 100 meters?

Ⓔ What was the highest score on the last exam?

10. Ms. Joanne did a survey about how many minutes her students spend to watch TV daily, and then she drew a line plot. What is the center of this data?

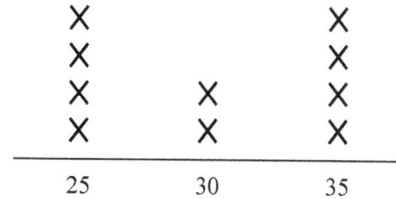

```
  X           X
  X           X
  X     X     X
  X     X     X
 _____
  25    30    35
```

Ⓐ 25　　　　Ⓑ 35　　　　Ⓒ 27.5

Ⓓ 30　　　　Ⓔ 32.5

11. What is the distance from the least value to the center of the data set,
{15, 18, 23, 26, 29, 34, 37}?

Ⓐ 22　　　　Ⓑ 11　　　　Ⓒ 14

Ⓓ 8　　　　Ⓔ 3

12. Ms. Kristy measured her students' height, and found that the mean of students' height is 53 inches and the variation (the difference between the mean and the greatest value) is 16 inches. If Andrew is the tallest student, what is his height?

Ⓐ 68 inches　　　　Ⓑ 69 inches

Ⓒ 77 inches　　　　Ⓓ 61 inches

Ⓔ 70 inches

13. Which of the following is the most appropriate name of the given figure?

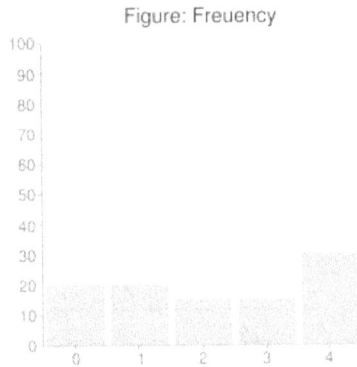

Figure: Freuency

Ⓐ Pie chart   Ⓑ Dot plot

Ⓒ Box plot   Ⓓ Histogram

Ⓔ Line chart

14. The graph below shows the number of new icecream sold in a icecream shop over five months. Based on the graph, what is the the best estimate of the average number of icecream sold each month?

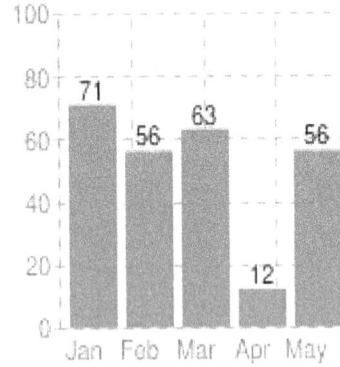

Ⓐ 260   Ⓑ 65   Ⓒ 208

Ⓓ 52   Ⓔ 57

15. The box plot for a data is given below.

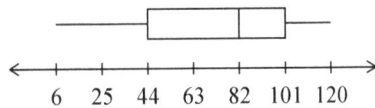

6   25   44   63   82   101   120

What is the range for the data?

Ⓐ 57   Ⓑ 82   Ⓒ 114

Ⓓ 76   Ⓔ 38

16. What is mean absolute deviation of the data set, $\{7, 12, 17, 22, 27\}$?

Ⓐ 8   Ⓑ 7   Ⓒ 5

Ⓓ 4   Ⓔ 6

Answer Keys (Scan item or test QR code to get detailed solution steps)

Skill Practice (6.RP.A.1)
1. $56:13$ 2. $8:5$ 3. 24 inches   4. $9:17$

Skill Practice (6.RP.A.2)
1. $\dfrac{5}{2}$ 2. 150   3. 21   4. 7

Skill Practice (6.RP.A.3)
1. 12   2. 90   3. 400   4. 48%

Skill Practice (6.RP.A.3.A)
1. 35   2. $1:8$ 3. $\dfrac{1}{5}$ 4. $\dfrac{1}{4}$

Skill Practice (6.RP.A.3.B)
1. 364   2. 36   3. $30   4. 40

Skill Practice (6.RP.A.3.C)
1. 6.5   2. 50%   3. 165   4. $16.66

Skill Practice (6.RP.A.3.D)
1. $1   2. $40.32   3. 32   4. 9

Comprehensive Exam (6.RP)
1. E   2. D   3. B   4. B
5. C   6. C   7. E   8. E
9. A   10. A   11. B   12. B
13. A   14. D   15. C   16. E

Skill Practice (6.NS.A.1)
1. 4   2. 69   3. $\dfrac{3}{7}$ 4. $\dfrac{12}{11}$

Skill Practice (6.NS.B.2)
1. A   2. 19   3. 4   4. 37

Skill Practice (6.NS.B.3)
1. 0.3   2. $11   3. 21.3   4. 3.4

Skill Practice (6.NS.B.4)
1. 16   2. 315   3. E   4. A

Skill Practice (6.NS.C.5)
1. $0   2. -13 3. -3 4. 0

Skill Practice (6.NS.C.6)

1. C **2.** $-5\frac{4}{5}$ **3.** S **4.** (3, 1)

Skill Practice (6.NS.C.6.A)
   1. S **2.** 39 **3.** P **4.** 0

Skill Practice (6.NS.C.6.B)
   1. (9, 4) **2.** Quadrant IV **3.** Quadrant IV **4.** Both

Skill Practice (6.NS.C.6.C)
   1. P **2.** R **3.** $4\frac{3}{4}$ **4.** (3, 1)

Skill Practice (6.NS.C.7)
   1. A **2.** A **3.** D **4.** -13

Skill Practice (6.NS.C.7.A)
   1. -6 **2.** E **3.** $-1 < 2$ **4.** C

Skill Practice (6.NS.C.7.B)
   1. D **2.** $a \geq 13$ **3.** $x < 26$ **4.** E

Skill Practice (6.NS.C.7.C)
   1. -10 **2.** 9.8 miles **3.** -3.4 **4.** B

Skill Practice (6.NS.C.7.D)
   1. Jane **2.** $t < -16$ **3.** E **4.** E

Skill Practice (6.NS.C.8)
   1. 22 **2.** 7 **3.** 20 **4.** 3

Comprehensive Exam (6.NS)
   1. E **2.** D **3.** A **4.** B
   5. C **6.** C **7.** D **8.** E
   9. A **10.** A **11.** B **12.** E
   13. A **14.** D **15.** C **16.** E

Skill Practice (6.EE.A.1)
   1. $5^3 + 3^3 + 6^2$ **2.** 26 **3.** $\frac{16}{25}$ **4.** -8

Skill Practice (6.EE.A.2)
   1. 42 **2.** $13^{o}C$ **3.** 37 **4.** 42

Skill Practice (6.EE.A.2.A)
   1. $2n - 12$ **2.** $35 - b$ **3.** $6(x + 9)$ **4.** $3y + 15$

Skill Practice (6.EE.A.2.B)
   1. -2 **2.** 3 **3.** -9 **4.** $t(5 + h)$

Skill Practice (6.EE.A.2.C)

1. 11　　　　**2.** $\dfrac{1}{64}$　　　　**3.** 8.5　　　　**4.** -6

Skill Practice (6.EE.A.3)
1. 9　　　　**2.** $7x + 8y$　　　　**3.** D　　　　**4.** $27x + 21$

Skill Practice (6.EE.A.4)
1. $47x - 15y$　　　**2.** $y^3 + y^2 + 4y$　　　**3.** $14x + 8$　　　**4.** $4y^3 + 5y^2 + 7y$

Skill Practice (6.EE.B.5)
1. B　　　　**2.** 7　　　　**3.** 5　　　　**4.** 2

Skill Practice (6.EE.B.6)
1. $\dfrac{1}{h}$　　　　**2.** $45h$　　　　**3.** $15 + c$　　　　**4.** $\dfrac{n}{2}$

Skill Practice (6.EE.B.7)
1. \$30　　　　**2.** \$3　　　　**3.** \$4.81　　　　**4.** 9

Skill Practice (6.EE.B.8)
1. $p \geq 2860$　　　**2.** A　　　**3.** $n > 100$　　　**4.** $x \leq 120$

Skill Practice (6.EE.C.9)
1. 13　　　**2.** $y = 3.89 \times x$　　　**3.** E　　　**4.** $C = 4n$

Comprehensive Exam (6.EE)
1. E　　　　**2.** D　　　　**3.** A　　　　**4.** A
5. C　　　　**6.** C　　　　**7.** E　　　　**8.** E
9. A　　　　**10.** A　　　　**11.** B　　　　**12.** D
13. A　　　　**14.** C　　　　**15.** C　　　　**16.** C

Skill Practice (6.G.A.1)
1. 54　　　　**2.** 30　　　　**3.** 47.5　　　　**4.** \$57

Skill Practice (6.G.A.2)
1. 48　　　　**2.** 125　　　　**3.** $\dfrac{765}{4}$　　　　**4.** 15

Skill Practice (6.G.A.3)
1. $(-3, -6)$　　　**2.** 126　　　**3.** 42　　　**4.** A

Skill Practice (6.G.A.4)
1. E　　　　**2.** 128　　　　**3.** 224　　　　**4.** 200

Comprehensive Exam (6.G)
1. E　　　　**2.** D　　　　**3.** B　　　　**4.** B
5. C　　　　**6.** C　　　　**7.** E　　　　**8.** E

Skill Practice (6.SP.A.1)
1. A　　　　**2.** E　　　　**3.** D　　　　**4.** A

Skill Practice (6.SP.A.2)
    1. 80   **2.** 55   **3.** E   **4.** D

Skill Practice (6.SP.A.3)
    1. E   **2.** 9   **3.** E   **4.** A

Skill Practice (6.SP.B.4)
    1. A   **2.** C   **3.** 23   **4.** A

Skill Practice (6.SP.B.5)
    1. 81   **2.** 27   **3.** 35   **4.** C

Skill Practice (6.SP.B.5.A)
    1. 22   **2.** 7   **3.** 125   **4.** 5

Skill Practice (6.SP.B.5.B)
    1. A   **2.** Day 4   **3.** D   **4.** B

Skill Practice (6.SP.B.5.C)
    1. 84   **2.** 65   **3.** 135   **4.** 75.5

Skill Practice (6.SP.B.5.D)
    1. B   **2.** D   **3.** D   **4.** E

Comprehensive Exam (6.SP)
    1. B   **2.** D   **3.** C   **4.** B
    5. C   **6.** B   **7.** E   **8.** C
    9. A   **10.** D   **11.** B   **12.** B
    13. D   **14.** D   **15.** C   **16.** E